MW00914185

The Y Tree

Kim Kennedy and Hector Lopez
Illustrated by Louie Lopez

This edition published by
Fabula Publishing
Redmond, WA

All rights reserved.

Printed in Canada by Friesens

Library of Congress Control Number: 2014903700
ISBN: 978-0-9915194-1-5

Text copyright © 2014 by Kim Kennedy and Hector Lopez
Illustration copyright © 2014 by Louie Lopez

www.thinkatree.com

To my mother, Charlotte Edgar, a beloved teacher who sparked the curiosity of generations of children– K.K.

To Isa, whose wondering nature stimulates my passion for exploration and discovery– H.L.

I see a Y in a tree, zippy, zoom, zing!

The tree is growing and now I see three.

1, 2, 3, 4, 5, 6, 7
This tree is in seven-Y heaven.

Look from bottom to top.
The Ys get smaller as you go up.

It grows some more. How many more?

Count them short and tall.
How many Ys are there in all?

There are 15 Ys in this tree.

Count them again.
Count them fast or count them slow.
Count them branch-by-branch or bunch-by-bunch.

1, add 2, add 4, add 8.

Fifteen total this does make.

It branches still more, step-by-step, fork-by-fork.
Can you foresee how many Ys there would be?

Aha! There are 31 Ys in our tree.

There are thick ones on the bottom

and thin ones on the top.

As long as the tree grows, the Ys won't stop.

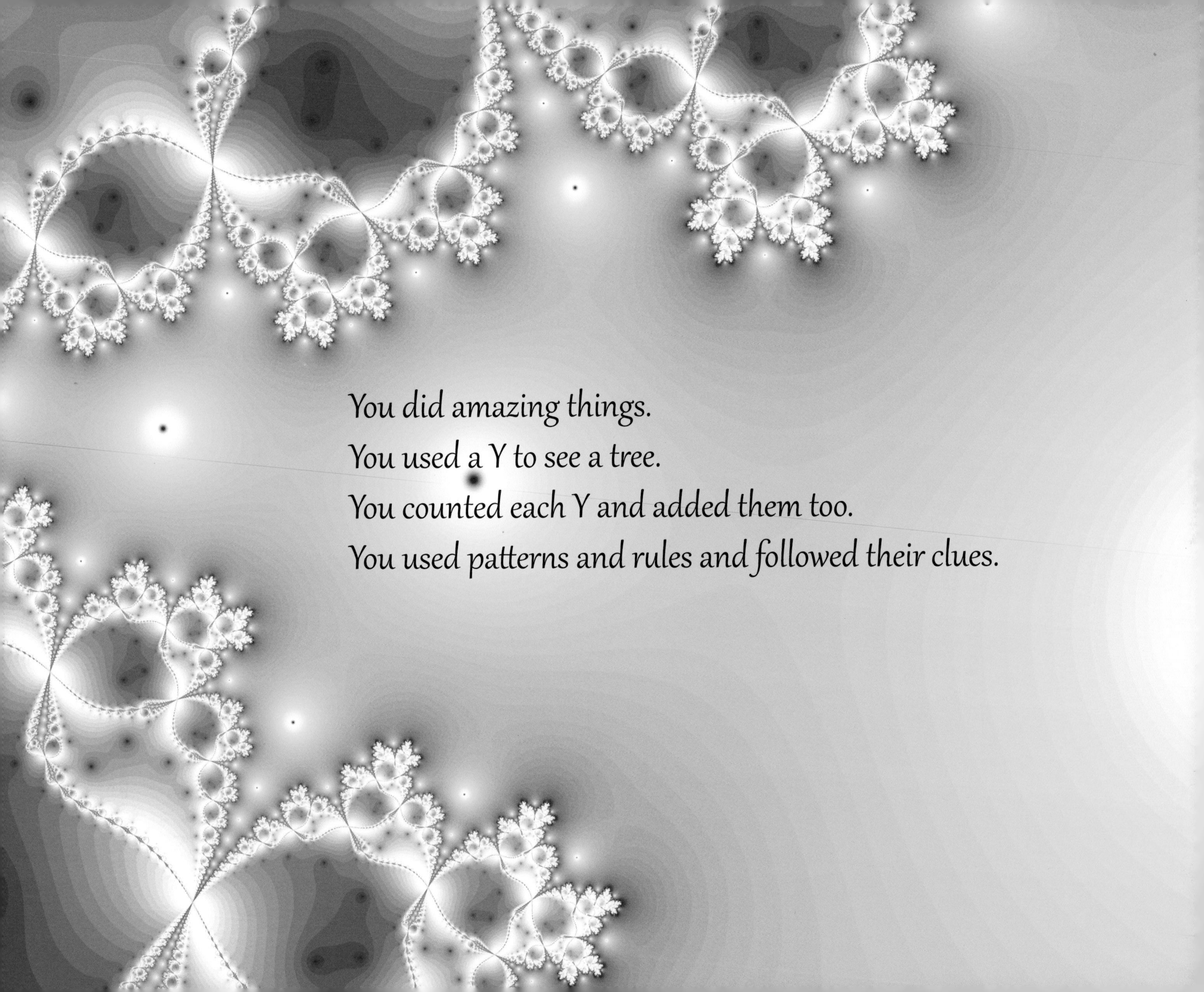

You did amazing things.
You used a Y to see a tree.
You counted each Y and added them too.
You used patterns and rules and followed their clues.

You worked your brain to think a tree:

a *fractal tree*, ZIPPY ZOOM ZING!

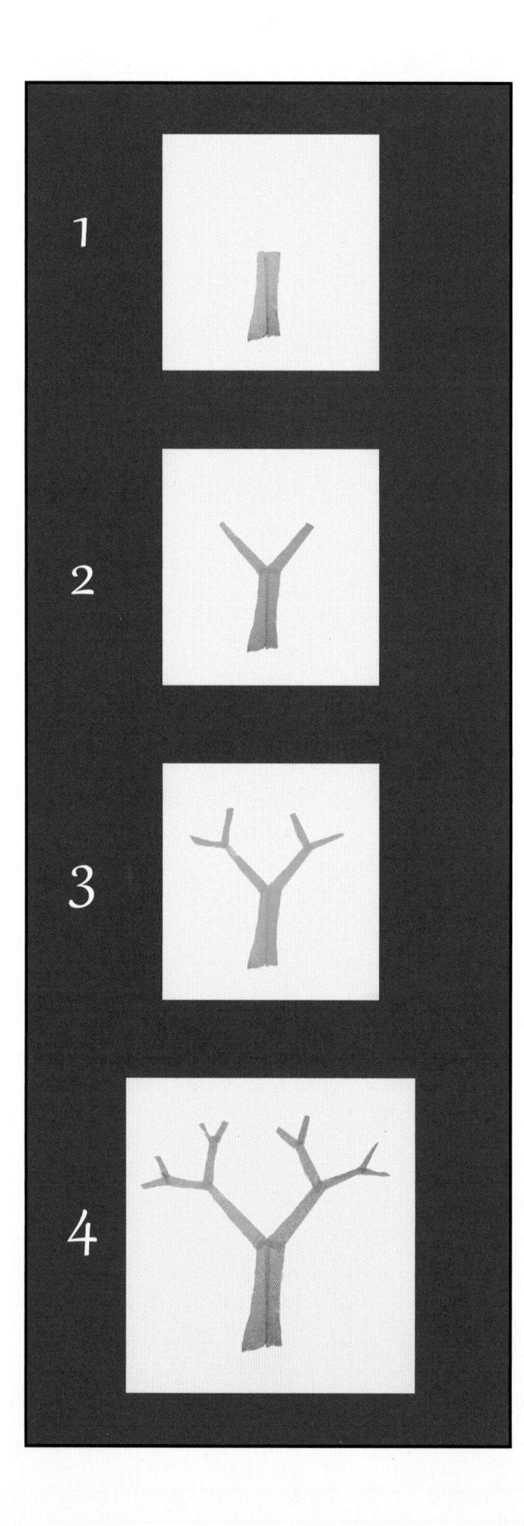

Make A Y Tree

This activity is a fun way for children to use rules to make a fractal tree. Building a Y tree helps children to see patterns and to connect these patterns to rules. You can create a Y tree using several different approaches. There are two simple rules regardless of the method that you use. Rule 1 is the binary rule: each branch divides into two. Rule 2 is the scale rule: the Ys get thinner and shorter at each branching step.

Cut tape or construction paper or draw on paper using paint, markers, or crayons. The following steps use tape to make a Y tree. Painter's tape works well because it is available in many colors and it is easy to remove from surfaces. Choose a window, glass door, or wall for your canvas.

1. Make the tree trunk. Cut 1 long piece of tape. Place this piece of tape near the bottom of your canvas. Tip: Use two long pieces of tape cut to the same size and place them side-by-side if you want a thicker trunk. Tearing the tape instead of cutting it will create a more natural tree effect. See the pictures on the left.
2. Make the first branching. Cut 2 pieces of tape to the same length. These pieces should be shorter and thinner than your trunk piece from step 1. Place one piece angled left at the top of your trunk. Place the other piece angled right at the top of your trunk. Take a look. You made a Y.
3. Make the second branching. Cut 4 pieces of tape to the same length. These pieces should be shorter and thinner than your branching pieces from step 2. Place one piece angled left at the top of your left branch. Place the other piece angled right at the top of your left branch. Repeat for the right branch. Take a look. You now have three Ys. Hint: Look for the fork (vertex) in each Y to make counting them easier.
4. Make the third branching. Cut 8 pieces of tape to the same length. They should be shorter and thinner than your branching pieces from step 3. Make Ys with these tape pieces. How many Ys do you see?

You can take it from here. Continue branching if you wish.

Create a different looking Y tree. Make the new branches shorter than the previous. Keep the width of the branches and trunk the same (i.e., the width of the tape).

Visit our website thinkatree.com for other Y tree activity ideas.

Nurturing Curiosity

We emphasize curiosity's role in motivating us to learn. Curiosity is connected to healthy habits of mind such as perseverance in solving difficult math problems. Curiosity inspires us to explore, observe, and learn. Depth and breadth of learning improve with our willingness to engage curiosities.

Moments of discovery and surprise are possible. You probably have heard others say "Aha!" when they suddenly see or understand something. These experiences are important because they are rewarding and enjoyable. Discovery can be fun and being able to understand what we could not previously understand helps us to learn and builds confidence. These sudden insights are fulfilling when we have been struggling and struggling to grasp something. Aha moments help us to remember what we have learned because such sudden discoveries tend to leave a lasting impression.

The Y Tree nurtures curiosity by encouraging shared discovery, using the mind, body, and senses to create opportunities for surprise. The book began with a spark of curiosity. Italian designer, Bruno Munari, explored the branching pattern of a tree in one of his drawings. This made us curious about how we could use tree branching and other fractal-like forms to teach math.

Cu•ri•os•i•ty, n.

A strong desire to know or learn

From the late Middle English: *from* Old French curiousete, *from* Latin curiositas, *from* curiosus

Oxford English Dictionary

Nurturing Curiosity #1

Find the dog in the picture below.

only the head

(Look at the branches for a hint.)

This Image is based on the photo "Female Dalmatian head shot" by Mllefantine at en.wikipedia.

Courtesy of Wikimedia Commons.

Visit thinkatree.com

for answers to the Nurturing Curiosity challenges and more fun fractal information.

About Fractals

This book introduces readers to fractal concepts. A fractal is a shape that is repeated over and over again. Fractals can be simple or complex. They can be created by following simple rules and steps. Most fractals today are generated and colorized using a computer program.

There are different types of fractals. The Y tree is a specific type called a binary fractal tree. It is one of the easiest fractals to make and can be drawn without the use of a computer.

One basic rule to draw a Y tree is to split a branch into two branches, starting at the trunk. The two-branch design is why we call it a binary tree. Both of these branches divide into two branches. This branching continues over and over. It is this binary branching rule that leads to the Y-shape pattern that is repeated for as long as you continue to draw branches. Another rule is that a new generation of Ys is smaller than the previous generation. This scaling pattern can be accomplished in many different ways depending on the rule that is followed. The particular rule followed dramatically influences the final shape and look of the Y tree.

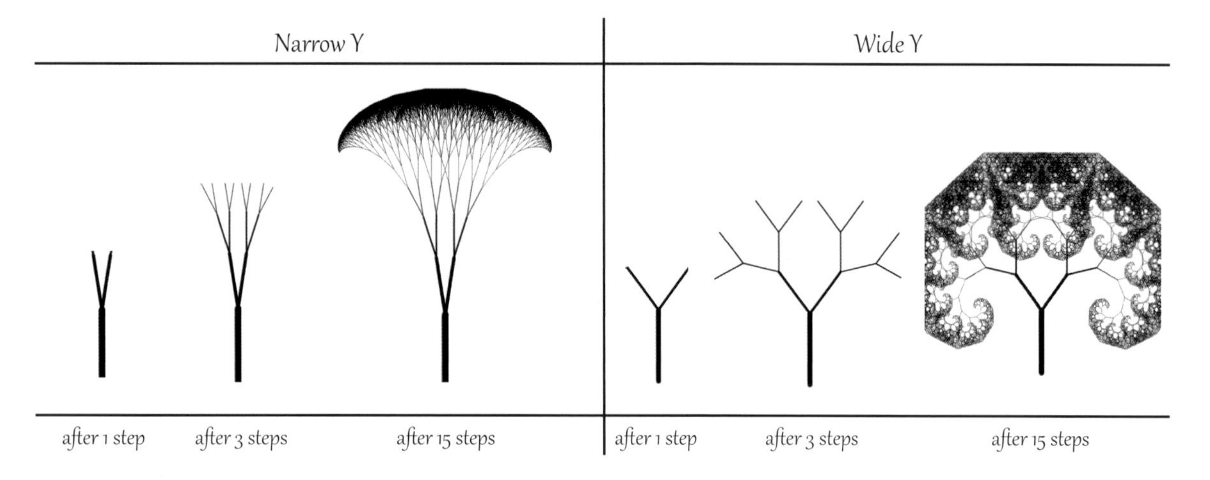

Another way of creating a fractal is by using escape-time algorithms. A computer program applies repeated mathematical equations and produces an image. The computer program will calculate where to draw a point and what color the point will be. The 'escape' condition is executed following a predetermined number of steps called iterations or until a condition is met. These mathematical steps result in beautiful, fractal art.

Two famous fractals are the Mandelbrot set and Julia set, named after mathematicians. All background illustrations in this book were generated using escape-time algorithms, most are derived from Mandelbrot and Julia sets.

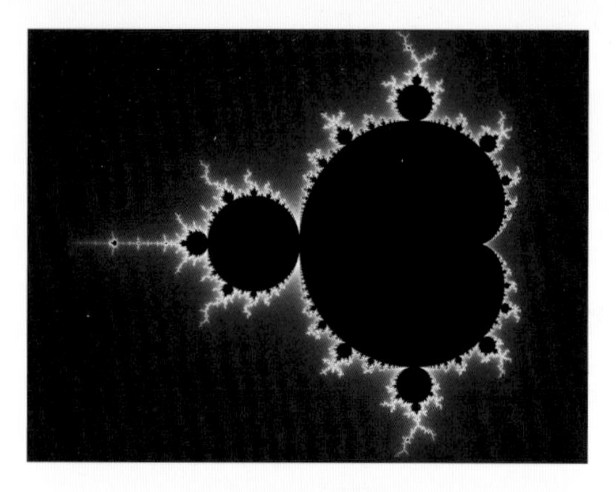

fractal Mandelbrot set

fractal Julia set

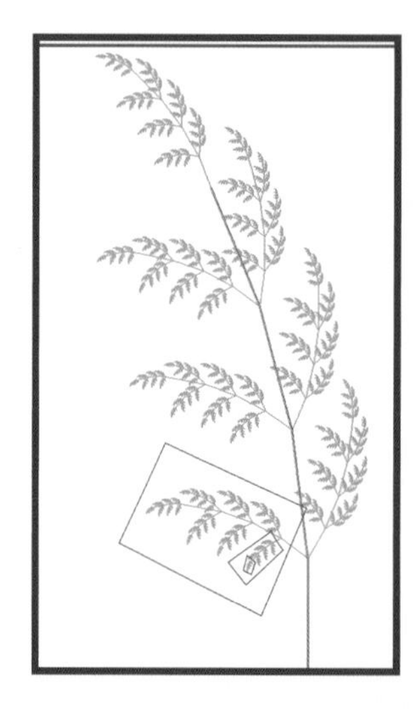

Self-similarity is a property of fractals that is the result of shapes repeating indefinitely and at different scales. The smallest part of a shape is similar to the whole shape.

The property of self-similarity is found in all illustrations in this book.

A self-similar "Fern Like Fractal" created with Fractal Grower 2010.03.

Nurturing Curiosity #2

Identify the shape repeated in each fractal presented within the book.

Other famous geometric fractals include:

Koch Snowflake

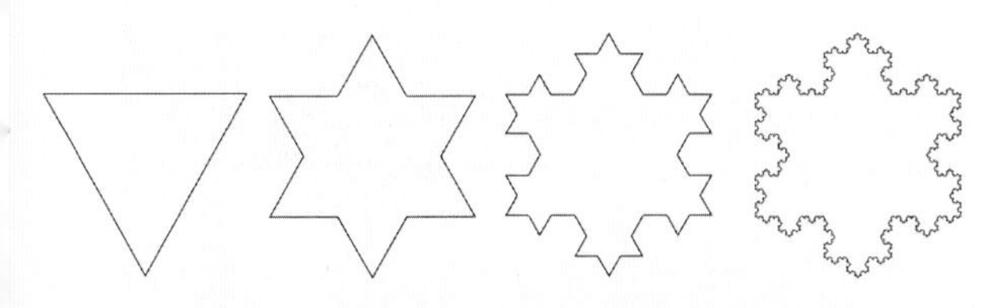

Sierpinski Gasket

Cutaway of a nautilus shell showing the chambers. "Nautilus CutawayLogarithmicSpiral" courtesy of Wikimedia Commons.

Fractals In Nature

You can find patterns that closely resemble fractals when you take a close look at the natural world. Throughout human history, artists have noticed and captured nature's patterns in paintings and sketches. For example, Leonardo da Vinci wrote a rule about branching.

Fractals in nature differ from mathematical fractals in an important way. In math, fractals can be on an infinite scale: a repeated, self-similar pattern can go on and on-into infinity. You will see the same pattern details at any magnification when you zoom in on mathematical fractals. Natural fractals have a limit and patterns repeat for a limited number of iterations.

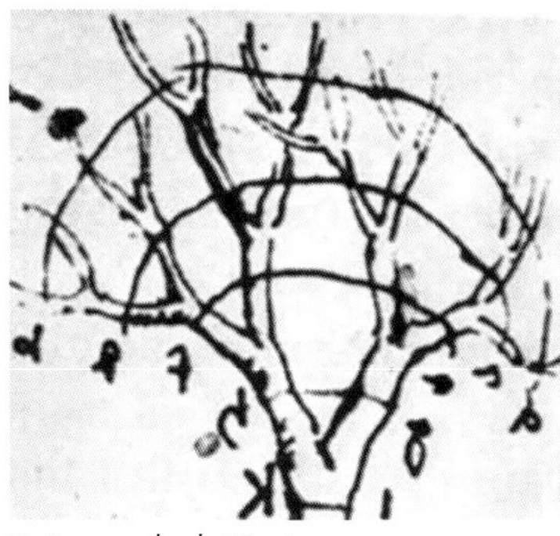

By Leonardo da Vinci

"Romanesco broccoli (3)" by Aurelien Guichard courtesy of Wikimedia Commons.

Galactic Spiral "UGC 12158" by ESA/Hubble & NASA

Nurturing Curiosity #3

Look around indoors or outdoors. Take pictures of self-similar patterns that you find. Share them with other readers by uploading your examples to our website at thinkatree.com.

Photo of a fern by Louie Lopez

Some Y Tree Math

The math in *The Y Tree* focuses on a simple counting problem—how many Ys are in the tree? The level of mathematics that can be explored ranges from preschool to secondary mathematics.

Preschool Mathematics. The first step in any counting problem is to identify the pattern to be counted; in this case it is a Y. Parents can help children identify the Y by outlining the Y with their finger, and holding the child's finger along the Y. As the tree "grows," it becomes harder and harder to find all the Ys. Having children point to as many Ys as they can locate is a great start. Mastery is achieved when children identify all the Ys in a tree. In addition, children can develop their one-to-one correspondence by counting the Ys. This is a foundational mathematical skill—the ability to match one object to one number.

Elementary Mathematics. The Y tree is a binary fractal tree. The Ys are created in a simple and systematic fashion. Step-by-step, each branch is split into two smaller branches. It is this recursive process that forms the Y tree. It is this recursive rule that makes it possible to derive the arithmetic series that determines the number of Ys in the tree.

A good way for children to collect, organize, and analyze information is to record data in a table. For the Y tree, you start in step 1 by seeing one Y. In the next iteration, step 2, you counted three Ys and so forth. The observations are tabulated as shown in the picture.

A number of patterns begin to emerge. For example, notice that the number of Ys in a Y tree is always an odd number. Another pattern occurs when you look at differences between two numbers from step-to-step. The difference between the number of Ys in iteration 1 and iteration 2 is 3-1=2. The difference between iteration 2 and iteration 3 is 7-3=4 and so forth. The difference between the number of Ys between two consecutive iterations is always an even number.

In fact, it is this pattern that often leads children to a series solution to the number of Ys in a binary fractal tree.

Iteration	Number of Ys	Series Solution
1	1	
2	3	1+2 = 3
3	7	1+2 +4= 7
4	15	1+2 +4+8 = 15

Nurturing Curiosity #4

Calculate the number of Ys for the 5th iteration using the next number of the series 1+2+4+8+?=?

Later Grades Mathematics. Further study of the table reveals other interesting solutions. For instance, two recursive formulas are evident. A recursive formula allows you to determine a value in a sequence by using values of a preceding sequence.

n	S_n = Sum	Recursive Solution
1	S_1	1
2	S_2	$S_1 + 2 = 3$
3	S_3	$S_2 + 4 = 7$
4	S_4	"$S_3 + 8 = 15$

For example, to calculate S_2 we need to know S_1 and to calculate S_3 we need to know the value for S_2, and so forth. We can write the recursion formula as

$$S_n = S_{n-1} + 2^{n-1}$$

with the initial condition that $S_0 = 0$. Where n is a positive integer greater than 1 and represents the iteration step. A different, but equivalent, recursive formula can be written as

$$S_n = 2S_{n-1} + 1.$$

Using both recursive formulas students can apply basic algebra to derive the general solution for the number of Ys in a binary tree,

$$S_n = 2^n - 1.$$

Nurturing Curiosity #5

How many Ys are in the binary tree with 11 iterations?